INTRODUCTION

Foundation Before Growth

Cultivation fails for one consistent reason: the foundation is treated as secondary.

Plants are often blamed. Weather is blamed. Genetics are blamed. Inputs are blamed. Rarely is the medium itself questioned beyond surface adjustments. This is backward. Growth is not the starting point. The medium is.

This book does not teach cultivation techniques. It does not provide schedules, formulas, or ratios. It does not attempt to optimize yield or speed. Those goals encourage fragility. Instead, this book focuses on **foundational truths that hold regardless of method, crop, or individual preference**.

Experience does not guarantee success. Repetition does not guarantee understanding. Equipment does not guarantee control. Only healthy systems endure correction.

Outdoor, living soil systems outperform artificial systems because they are **self-regulating, buffered, and biologically aligned**. Artificial systems rely on substitution. Living systems rely on relationship. Substitution always collapses under stress. Relationship adapts.

Everything that follows is open-source knowledge. None of it is proprietary. None of it requires belief. These principles function whether they are acknowledged or not.

The conclusion is simple and unavoidable:

If the foundation stays healthy, growth follows. If the foundation degrades, nothing compensates.

CULTIVATING REALITY

Cultivating Reality

GARY HAYWOOD

I
The Illusion of Control

M odern cultivation is dominated by control logic.

Inputs are isolated. Variables are measured. Environments are stabilized. This approach is attractive because it appears scientific. It produces visible, repeatable outcomes under narrow conditions. The illusion is not that control fails. The illusion is that control scales.

.

Artificial systems work by replacing natural processes with direct intervention. Light replaces the sun. Nutrients replace soil biology. Airflow replaces wind. Containers replace ecosystems. Each substitution functions independently. Together, they create dependence. The more substitutions introduced, the more fragile the system becomes. Control-based systems demand precision because they lack buffers. When a system cannot self-correct, every deviation becomes a failure event. This creates a false sense of mastery. The operator is constantly "fixing" outcomes that the system itself created.

Living systems behave differently. Outdoor environments introduce variability by default. Temperature fluctuates. Moisture changes. Light shifts. Biology responds dynamically. Stress is distributed rather than concentrated. Correction occurs continuously, not episodically. This is not romanticism. It is structure. Artificial systems remove friction. Living systems distribute it. Friction is not an enemy of growth. It is a conditioning force. Plants grown under stable, artificial conditions adapt to support structures, not environments.

Remove one support, and the plant falters. Plants grown in variable outdoor conditions adapt to the environment itself. Sunlight is not merely brighter than artificial light. It is complete, rhythmic, and dynamic. It changes angle, intensity, and spectrum throughout the day and season. This variability forces adaptation at every level of plant function. Artificial light produces growth. Sunlight produces resilience.

Attempts to replicate nature indoors inevitably simplify it. Simplification always removes redundancy. Redundancy is what prevents collapse. The result is predictable: artificial systems plateau. They reach a performance ceiling that cannot be crossed without increasing complexity. Increased complexity increases failure probability. The system becomes brittle.

Outdoor, soil-based systems scale differently. They do not require constant correction because the system itself performs correction. When something is off, biology responds before collapse occurs, if the foundation is alive.

Control feels productive. Alignment produces results. The purpose of cultivation is not to dominate the environment.

It is to **build a foundation that functions without constant intervention**. That foundation is soil.

II
Soil as a Living System

S oil is not dirt.

Dirt is inert. Soil is organized life. A healthy medium is not defined by nutrient content alone. Nutrients without biology are raw materials without logistics. Biology governs access, timing, buffering, and correction. Remove biology, and nutrients become liabilities rather than assets. Soil functions as infrastructure. It regulates moisture. It moderates temperature. It buffers chemical imbalance. It houses microbial populations that convert unavailable compounds into usable forms. It stores excess rather than allowing accumulation. It releases slowly rather than dumping abruptly.

When soil is treated as a container, these functions are lost. When soil is treated as a system, these functions dominate outcomes. The primary actors in living soil are bacteria and fungi. Bacteria specialize in rapid cycling. They break down simple compounds, respond quickly to moisture and temperature changes, and drive early-stage nutrient availability. Fungi specialize in transport, structure, and long-term stability. Mycorrhizal networks extend root systems, regulate exchange, and improve resilience under stress.

Balance matters more than volume. Excess bacteria without fungal structure leads to instability. Excess fungi without bacterial cycling leads to stagnation. Healthy soil contains both, operating within a stable physical structure.

Structure is not cosmetic. Aggregation determines oxygen availability, water movement, and root penetration. Compacted soil suffocates biology. Overly loose soil dries too quickly. Healthy soil holds shape, drains excess, and retains moisture.

Biology follows structure. Inputs do not create life. Conditions allow it. This is where most failures occur. Inputs are added without regard for existing balance. Sugars are added when bacterial populations are already dominant. Nitrogen is added when mineral imbalance is the real issue. Amendments are layered without allowing time for integration.

Time is not optional. Biological systems operate on biological timelines. Cold soil slows activity regardless of air temperature. Dormant biology cannot be forced awake without damage. Attempts to accelerate correction usually extend dysfunction.

Soil absorbs mistakes when it is alive. Dead media amplifies them. This buffering capacity is the defining advantage of outdoor, living soil systems. Errors occur, but collapse is rare. Correction happens continuously at low intensity rather than episodically at high intensity. Healthy soil does not require constant feeding.

It requires **maintenance of conditions**.

Those conditions are simple:
Adequate organic matter
Stable structure
Balanced moisture
Mineral sufficiency without excess
Active, diverse biology

When these are present, plants do not need to be managed aggressively. They respond to the system provided.

The medium determines the ceiling. The plant expresses what the medium allows.

III
Inputs, Balance, and Repeatable Outcomes

I nputs are not solutions.

They are signals. Every addition to soil shifts biological balance. Some shifts are corrective. Others are destabilizing. The difference is not the ingredient itself, but context.

This is why recipes fail. Recipes assume uniform starting conditions. Living systems never provide them. Two soils can receive identical inputs and diverge dramatically. The outcome depends on existing biology, structure, mineral profile, moisture, and temperature. Open source cultivation acknowledges this reality. It focuses on function rather than formula.

Compost teas are not fertilizers. They are biological inoculations. Their value lies in microbial diversity and activity, not nutrient strength. Aged manures are not nitrogen sources alone. Properly aged material contributes organic matter, microbial diversity, and slow-release nutrition. Improperly aged material damages biology and structure.

Worm castings function as biological concentrates. They improve structure, microbial diversity, and nutrient availability simultaneously. Their effectiveness depends on moderation.

Unsulfured molasses is not plant food. It is microbial fuel. Applied with intent, it supports biological recovery.

Lactic acid bacteria influence microbial balance and pathogen suppression. They are corrective tools, not routine additives. Repeated use without assessment degrades diversity.

Minerals matter, but accumulation matters more. Deficiencies are often misdiagnosed. What appears as lack is frequently imbalance. Adding more of a single element without correcting ratios creates lockout rather than resolution. Soil stores excess. Plants suffer the consequence.

The principle is consistent: **Feed the system, not the symptom.**

Observation replaces instruction. Healthy soil announces itself clearly:

Water infiltrates evenly
Soil smells clean and earthy
Structure holds without compaction
Roots explore freely
Growth resumes without forcing

When these signals are present, restraint is required. Intervention at this stage introduces instability. When these signals are absent, correction must be structural, not cosmetic. This is where patience becomes operational, not philosophical. Biological correction does not respond to urgency. Attempts to rush recovery usually reset progress. The outcome, however, is repeatable. Across climates, crops, and individual preferences, the same result appears when the foundation is respected: **a healthy medium, rich in micronutrients and biological life, capable of sustaining growth without constant correction.**

Ingredients will differ. Soil composition will differ. Climate will differ. The end state converges.

Healthy systems look the same because function converges even when inputs differ.

This is the defining advantage of living soil cultivation.

It is not dependent on the cultivator's experience, intuition, or precision. It depends on adherence to foundational truths.

Keep the foundation healthy. Growth follows.

Ignore the foundation. Everything else becomes effort.

Foundation is not Optional

Cultivation does not reward control. It rewards alignment.

Artificial systems create performance under constraint. Living systems create resilience under pressure.

One requires constant management. The other corrects continuously.

This book does not argue against technology. It argues against mistaking substitution for structure.

The medium determines the outcome.
Biology determines the medium.
Balance determines biology.

These truths do not change with trends, equipment, or experience. They apply universally because they are structural, not instructional.

Every individual will choose different inputs.
Every soil will differ.
Every environment will impose unique constraints.
None of that alters the conclusion.

If the foundation stays healthy, growth is inevitable. If the foundation degrades, intervention multiplies.

Cultivating reality means accepting this hierarchy and working within it.   There are no shortcuts around the foundation.

Operating in Reality

Every book I've written points to the same conclusion, even when the subject changes: outcomes are not created by intention, belief, or appearance. They are produced by systems that function under real conditions.

Cultivating Reality fits squarely within that framework.

This book is not about nostalgia or rejection of technology. It is about recognizing when tools become substitutes for understanding. Soil does not need to be controlled, it needs to be respected. Growth does not need to be forced, it needs to be supported.

Reality does not reward aesthetics. It rewards alignment.

When you build systems, whether personal, financial, or biological, that operate in truth, results follow without constant intervention. That is the lesson across all my work, and it is the foundation of *Cultivating Reality*.

Operate accordingly.